Abdelhafid Mimouni

Cytochrome P450: Exploration of Metabolic Enzymes

Abdelhafid Mimouni

Cytochrome P450: Exploration of Metabolic Enzymes

ScienciaScripts

Imprint

Any brand names and product names mentioned in this book are subject to trademark, brand or patent protection and are trademarks or registered trademarks of their respective holders. The use of brand names, product names, common names, trade names, product descriptions etc. even without a particular marking in this work is in no way to be construed to mean that such names may be regarded as unrestricted in respect of trademark and brand protection legislation and could thus be used by anyone.

Cover image: www.ingimage.com

This book is a translation from the original published under ISBN 978-620-6-71951-9.

Publisher:
Sciencia Scripts
is a trademark of
Dodo Books Indian Ocean Ltd. and OmniScriptum S.R.L publishing group

120 High Road, East Finchley, London, N2 9ED, United Kingdom
Str. Armeneasca 28/1, office 1, Chisinau MD-2012, Republic of Moldova, Europe
Printed at: see last page
ISBN: 978-620-7-97895-3

AUTHOR

Dr. Abdelhafid Mimouni is an independent researcher specialising in the chemistry of bioinorganic systems. He has extensive expertise in macromolecular synthesis and characterisation. He obtained his doctorate in chemistry from the University of Paris XII in 1997 and a Diplôme des études approfondies in bioinorganic systems from the University of Paris XI in 1993.

BOOK SUMMARY

This book explores in depth the essential role of cytochromes P450 (CYP), a family of key enzymes in the metabolism of drugs, lipids and steroids. We trace their history, their unique molecular structure and their biological functions, including their ability to catalyse complex oxidation reactions. The interactions of cytochromes P450 with other enzymes and metals are examined, highlighting their importance in metabolic pathways. The disorders associated with dysfunction of these enzymes and their impact on pharmacokinetics are discussed, as are their applications in biotechnology, such as biocatalysis and bioremediation. The book concludes with a reflection on future prospects, highlighting the challenges and opportunities for bioinorganics research, with potential implications for the therapeutic and environmental fields.

TABLE OF CONTENTS

CHAPTER 1

INTRODUCTION TO CYTOCHROME P450

History

Cytochromes P450 (CYP) are essential enzymes involved in a multitude of biological processes, including the metabolism of drugs, hormones and lipid compounds. The discovery of cytochromes P450 dates back to the 1950s, when researchers observed specific heme pigments in liver tissue. These pigments were identified as haemoproteins responsible for the transformation of various substrates. Cytochrome P450s were first identified by David R. Williams and his colleagues in the early 1960s. Williams discovered that these enzymes were responsible for the oxidation of organic compounds, a crucial process in the metabolism of drugs and toxins. The name "P450" comes from the characteristic colour of the proteins when they are reduced and exposed to light at a wavelength of 450 nm. Research into cytochromes P450 has evolved considerably since their initial discovery. In the 1970s and 1980s, advances in chromatography and spectroscopy techniques enabled more detailed characterisation of these enzymes. Researchers discovered that cytochromes P450 had a heme structure and played a crucial role in substrate oxidation reactions. Initial sequencing studies have revealed the diversity of cytochrome P450 families, each with specific substrates and functions. The evolution of cytochromes P450 as a research subject has accelerated with the development of molecular biology and DNA sequencing techniques. The mapping of the human genome and other mammalian genomes has made it possible to identify and characterise a large number of genes coding for P450 cytochromes. Today, more than 50 families of cytochromes P450 have been described, and their role in various physiological and pathological processes is well established.

Structure and Function

Cytochromes P450 are enzymes of the haemoprotein family, characterised by the presence of a haem group in their structure. The heme group consists of an iron atom linked to a porphyrin ring, which is essential for enzymatic function. This structure gives cytochromes P450 their unique ability to catalyse oxidation reactions on a wide variety of substrates. The structure of cytochromes P450 comprises several functional domains:

- **The heme domain**: This is the part of the enzyme where the catalytic reaction takes place. The iron in the heme group plays a crucial role in transferring the electrons needed to oxidise the substrates.

- **Substrate Binding Domain**: This region is responsible for interaction with the enzyme's specific substrate. Substrate recognition is often modulated by conformational changes in the enzyme structure.

- **The Electron Transfer Domain**: This domain interacts with reducing proteins to supply the electrons needed for the catalytic reaction.

Cytochromes P450 are involved in various types of biochemical reactions, including :

• **Hydroxylation**: Introduction of a hydroxyl group (-OH) into the substrate.

• **Oxidation**: Addition of an oxygen atom to the substrate, often in the form of an oxide or hydroxide.

• **Demethylation**: Removal of methyl groups (-CH3) from substrates. These reactions play a fundamental role in the metabolism of drugs, lipids and hormones. By metabolising drugs, cytochromes P450 facilitate their elimination from the body by making them more water-soluble. They are also involved in the synthesis and degradation of steroid hormones and in the biotransformation of lipids.

Cytochrome P450s vary widely in their specificity and affinity for substrates, which is determined by their structure. three-dimensional. This diversity enables them to catalyse a wide range of reactions, making these enzymes essential for maintaining body homeostasis and responding to exogenous substances.

Synthesis of cytochromes P450: Processes of transcription, translation and post-translational modifications, including the maturation and assembly of cytochromes P450.

Cytochrome P450 (CYP) synthesis is a complex process that takes place mainly in the endoplasmic reticulum of eukaryotic cells. The process comprises several key stages: transcription, translation, post-translational modifications and protein maturation. Here is a detailed overview of each stage:

1. Transcription

Cytochrome P450 synthesis begins with the transcription of the corresponding gene into messenger RNA (mRNA). This stage takes place in the nucleus of the cell, where RNA polymerase II copies the gene into precursor mRNA.

• **Location**: Cell nucleus

• **Mechanism**: DNA is transcribed into mRNA by RNA polymerase II. The pre-messenger RNA then undergoes post-transcriptional modifications such as the addition of a 5' cap and a poly-A tail, as well as intron splicing to produce a mature mRNA.

2. Translation

Once mature mRNA has been exported from the nucleus to the cytoplasm, it is translated into protein by ribosomes. Being a membrane protein, cytochrome P450 benefits from a special translation on the rough endoplasmic reticulum.

• **Location**: Cytoplasm (linked to the endoplasmic reticulum)

• **Mechanism**: Ribosomes attached to the rough endoplasmic reticulum translate mRNA into a polypeptide chain. This precursor chain is inserted into the endoplasmic reticulum during translation.

3. Translocation and Post-Translational Modifications

The newly synthesised polypeptide chain is then translocated to the endoplasmic reticulum where it undergoes the various post-translational modifications required for maturation.

- **Location**: Endoplasmic reticulum

- **Mechanism**:

o **Insertion into the membrane**: The precursor protein is inserted into the membrane of the endoplasmic reticulum.

o **Cleavage**: The precursor protein is cleaved to form mature cytochrome P450.

o **Addition of the heme group**: The heme group, essential for enzymatic activity, is incorporated into the protein at the level of the endoplasmic reticulum.

o **Folding and assembly**: The protein folds correctly and assembles into a functional structure in the endoplasmic reticulum.

4. Transport and Function

After maturation, cytochrome P450 is transported to its functional sites. Primarily located in the endoplasmic reticulum, it can also be found in the mitochondrial membrane in certain cell types.

- **Location**: Endoplasmic reticulum (mainly), mitochondrial membranes in certain cell types

- **Function**: Cytochrome P450 catalyses oxidation reactions on various substrates, including drugs, toxins and endogenous molecules.

Summary

1. **Transcription**: In the nucleus, the cytochrome P450 gene is transcribed into mRNA.

2. **Translation**: In the cytoplasm, the mRNA is translated into a polypeptide chain by ribosomes attached to the endoplasmic reticulum.

3. **Translocation and modifications** : In the endoplasmic reticulum, the precursor protein is inserted, cleaved and modified (addition of the heme group) to become a mature cytochrome P450.

4. **Transport and function**: Mature cytochrome P450 is transported to the endoplasmic reticulum or mitochondrial membrane, where it carries out its enzymatic functions.

This process ensures the production and maturation of cytochromes P450, enabling them to carry out their roles effectively in the metabolism of biological substrates.

CHAPTER 2

BIOLOGICAL ROLE OF CYTOCHROMES P450

Main function

Cytochromes P450 play a central role in the metabolism of drugs, lipids and steroids. These enzymes are present in various tissues, but are particularly abundant in the liver, where they facilitate the detoxification and elimination of exogenous and endogenous substances.

Metabolism of Drugs Cytochromes P450 are responsible for the biotransformation of many drugs, often by oxidation. This reaction modifies the chemical structure of the drugs, generally to make them more water-soluble and therefore easier to excrete. For example, methadone, an opioid analgesic, is metabolised mainly by the CYP3A4 and CYP2B6 isoenzymes. The diversity of P450 isoenzymes in the liver enables a wide variety of drugs to be metabolised quickly and efficiently, but it can also lead to complex drug interactions. Genetic variations in the genes encoding these enzymes can influence the rate of metabolism, leading to varying effects in response to treatment.

Lipid metabolism

Cytochromes P450 also play a crucial role in lipid metabolism, particularly in the biotransformation of fatty acids and lipoproteins. These reactions are essential for regulating lipid levels in the body. These enzymes are also involved in the synthesis of bioactive molecules such as eicosanoids, which are implicated in inflammatory and immune processes. For example, the CYP4A and CYP4F enzymes are involved in the hydroxylation of fatty acids, thus influencing the composition of lipids and their distribution in the body.

Steroid metabolism

Cytochromes P450 are also involved in the metabolism of steroid hormones, such as corticosteroids, androgens and oestrogens. These enzymes are responsible for converting hormones into their active or inactive forms, thereby regulating various aspects of hormonal physiology. For example, the CYP17 and CYP19 enzymes are crucial in the biosynthesis of sex steroids. Disruption of these enzymes can lead to hormonal imbalances and endocrine disorders.

Mechanism of Action

Cytochrome P450s catalyse oxidation reactions using a complex mechanism involving several steps and interaction with various cofactors.

Catalytic Reaction

The catalytic reaction typical of cytochromes P450 begins with the association of the substrate with the enzyme's active site. The heme group of cytochrome P450, which contains an iron atom, plays a crucial role in catalysis. Molecular oxygen is activated by the iron in the heme group, and the electrons required for the reaction are supplied by reducing proteins, such as NADPH-cytochrome P450 reductase.

Oxidation mechanism

The oxidation mechanism of cytochromes P450 is often described as a multi-step cycle:

1. **Substrate binding** : The substrate binds to the enzyme's active site, modifying its conformation.

2. **Oxygen activation**: Molecular oxygen binds to the iron in the heme group and is activated.

3. **Electron transfer** : Electrons are transferred from NADPH to cytochrome P450, enabling oxygen to be activated.

4. **Formation of the Intermediate Complex**: Activated oxygen combines with the substrate, forming an intermediate complex.

5. **Product release** : The oxidised product is released, and the enzyme is ready to catalyse a new reaction.

Interactions with Cofactors

NADPH-cytochrome P450 reductase is the main cofactor that supplies the electrons required for catalytic reactions. Interactions between cytochrome P450 and this reductase are essential for the proper functioning of cytochrome P450. Other cofactors and regulators can also influence enzyme activity, including membrane lipids and allosteric modulators.

Regulation of Expression

Cytochrome P450 expression is finely regulated at transcriptional and post-transcriptional levels, enabling dynamic adaptation to metabolic needs and environmental stimuli.

Transcriptional regulation

The expression of genes encoding cytochromes P450 is regulated by specific transcription factors. For example, nuclear receptors such as the xenobiotic receptor (CAR) and the oestrogen receptor (ER) can modulate the expression of certain P450 cytochromes in response to chemical substances or hormones. These receptors bind to specific response elements in the promoters of CYP genes, thereby influencing their transcription.

Post-transcriptional regulation

After transcription, cytochrome P450 expression can also be regulated by post-transcriptional mechanisms, such as messenger RNA (mRNA) degradation and translation regulation. MicroRNAs (miRNAs) play an important role in this process by interfering with the stability of cytochrome P450 mRNAs or inhibiting their translation.

Influence of Environmental Factors

Cytochrome P450 expression can also be modified by environmental factors such as diet, infection and exposure to toxins. For example, chronic exposure to drugs or toxins can induce the expression of certain P450 isoenzymes to increase their detoxification capacity. Conversely, pathological conditions such as liver failure can reduce the expression of cytochromes P450, thereby affecting drug metabolism.

CHAPTER 3

INTERACTION WITH OTHER ENZYMES AND METALS

Enzyme interactions

Cytochrome P450s (CYPs) play a crucial role in various complex enzymatic networks, interacting with many other enzymes in metabolic pathways. Their function and efficiency are often the result of their collaboration with other proteins and enzyme systems.

Interactions with Reducing Proteins

Cytochrome P450s require reducing proteins to supply the electrons needed for the oxidation reaction. NADPH-cytochrome P450 reductase is the main reductase involved in this process. This enzyme transfers electrons from NADPH to cytochrome P450, facilitating oxygen activation and catalysis of the reaction. Interactions between cytochrome P450 and NADPH-cytochrome P450 reductase are crucial to the proper functioning of cytochrome P450.

Collaboration with Phase I and II Enzymes

Cytochrome P450s play a central role in phase I of xenobiotic metabolism, where they chemically modify compounds by oxidation. These intermediates are then often conjugated with hydrophilic groups by phase II enzymes, such as glutathione S-transferases (GST) and UDP-glucuronosyltransferases (UGT). These conjugations increase the solubility of metabolites, facilitating their excretion. Interactions between cytochromes P450 and phase II enzymes are therefore essential for the effective detoxification of chemical substances.

Interactions with Lipid Metabolism Enzymes

The cytochromes P450 involved in lipid metabolism, such as CYP4A and CYP4F, interact with other enzymes in the eicosanoid pathway. These enzymes modify polyunsaturated fatty acids to produce bioactive mediators such as prostaglandins and leukotrienes, which regulate various physiological and pathological processes. The interactions between cytochromes P450 and these enzymes are crucial for maintaining lipid balance and the inflammatory response.

Interactions in Steroid Biosynthesis

In the biosynthesis of steroid hormones, cytochromes P450, such as CYP17 and CYP19, interact with other enzymes and hormone regulators. For example, CYP17 catalyses essential reactions in the biosynthetic pathway of androgens and corticosteroids, while CYP19 is involved in the conversion of androgens to oestrogens. These interactions coordinate hormone production and influence hormone levels in the body.

Role of Metals

Cytochromes P450 are haemoproteins whose function is highly dependent on metals, particularly iron. However, other metals also play important roles in the function and regulation of cytochromes P450.

Iron in the Heme Centre

The heme centre, consisting of an iron atom bound to a porphyrin, is the key element in the function of cytochromes P450. The iron in the heme group is responsible for binding and activating molecular oxygen. In the mechanism of action of cytochromes P450, molecular oxygen binds to the iron of the heme group and is activated, allowing oxidation of the substrate. Iron also plays a role in stabilising intermediate states during the catalytic reaction.

Other Metal Cofactors

In addition to iron, other metals can influence the function of cytochromes P450:

• **Zinc**: Some P450 cytochromes contain zinc ions in their structure, which can stabilise the conformation of the enzyme or modulate its activity.

• **Magnesium**: Magnesium is sometimes involved in substrate binding processes or in the regulation of enzyme activity.

• **Calcium**: Calcium can influence the regulation of cytochromes P450 by modulating their interaction with other proteins or by affecting their conformational structure.

Metals as Inhibitors or Inducers

Certain metals can also influence the activity of cytochromes P450 as inhibitors or inducers. For example, copper and cobalt ions can interfere with the active site of P450 cytochromes, altering their ability to catalyse oxidation reactions. Other metals can induce the expression of certain P450 isoenzymes, thus affecting the metabolism of drugs and toxins.

Influence of Metals on the Regulation of P450 Cytochromes

The presence and concentration of various metals in the body can influence the regulation of cytochromes P450. For example, high levels of certain heavy metals can induce changes in the expression of P450 cytochromes, leading to alterations in the metabolism of xenobiotics. This complex interaction between metals and cytochromes P450 may have important implications for pharmacology and toxicology.

Classification of P450 cytochromes

Cytochromes P450 are classified into families and subfamilies according to the similarity of their amino acid sequences and their specific functions. Nomenclature is established to reflect this classification, making it easier to understand and communicate the various types of P450 cytochromes.

Nomenclature System

Cytochromes P450 are classified on the basis of amino acid sequence similarity:

- **Family**: Cytochromes P450 belonging to the same family have a sequence similarity of at least 40%.
- **Subfamily**: Cytochromes P450 within the same family with at least 55% sequence similarity are grouped together in the same subfamily.
- **Isoform**: The different isoforms within the same subfamily are identified by a number.

Main families of P450 cytochromes

• **CYP1 family** :

o**Subfamilies**: CYP1A, CYP1B

o**Examples**: CYP1A1, CYP1A2, CYP1B1

o**Roles**: Metabolism ofxenobiotics, activation of carcinogens,

metabolism of steroid hormones.

• **CYP2 family** :

o**Subfamilies**: CYP2A, CYP2B, CYP2C, CYP2D, CYP2E

o**Examples**: CYP2C19, CYP2D6, CYP2E1

o**Roles**: Metabolism of drugs, toxins, endogenous compounds such as

hormones and lipids.

• **CYP3 family** :

o**Subfamilies**: CYP3A

o**Examples**: CYP3A4, CYP3A5, CYP3A7

o**Roles** Metabolism of many drugs, interactions, steroid

biotransformation.

- **CYP4 family** :

o **Subfamilies**: CYP4A, CYP4B, CYP4F

o **Examples**: CYP4A11, CYP4F2

o **Roles**: Metabolism of fatty acids, production of eicosanoids.

- **CYP5 to CYP8 family** :

o These families are less well studied but play roles in various specific biochemical reactions.

- **CYP11 family**:

o **Subfamilies**: CYP11A, CYP11B

o **Examples**: CYP11A1, CYP11B1

o **Roles**: Biosynthesis of steroids, regulation of adrenal hormones.

- **CYP17 family** :

o **Examples**: CYP17A1

o **Roles**: Steroid metabolism, biosynthesis of androgens and corticosteroids.

- **CYP19 family** :

o **Examples**: CYP19A1

o **Roles**: Aromatase, conversion of androgens into oestrogens.

Examples of notable isoforms

• **CYP1A2**: Metabolises drugs such as caffeine and certain psychotropic drugs. Involved in the bioactivation of certain carcinogens.

• **CYP2D6**: Responsible for the metabolism of many psychotropic and analgesic drugs, with considerable genetic variability influencing enzymatic activity.

• **CYP3A4**: Most abundant in the liver, metabolises around 50% of common drugs. Interacts with many drugs, influencing pharmacokinetics and drug interactions.

CHAPTER 4

CYTOCHROME-RELATED DISEASES AND DYSFUNCTIONS

P450

Cytochromes P450 (CYP) play an essential role in the detoxification of harmful substances and the metabolism of biological compounds. Their dysfunction or abnormalities can lead to various metabolic disorders and affect the efficiency of biological processes. This chapter explores the diseases associated with cytochromes P450, their interactions with other substances, and their impact on the degradation of toxins and drugs.

Metabolic diseases

Cytochrome P450 enzymes are crucial for breaking down toxic substances and foreign compounds in the body. Mutations or abnormalities in the genes encoding these enzymes can lead to serious metabolic disorders. These pathologies include :

- **Cytochrome P450 deficiency syndromes**: These syndromes result from genetic mutations affecting the expression or function of cytochrome P450s. For example, deficiency of CYP21A2, which plays a role in the synthesis of corticosteroids, can lead to severe endocrine disorders such as congenital adrenal hyperplasia. Deficiencies in other CYP family enzymes can also lead to disorders in lipid, steroid and drug metabolism.

• **Lipid metabolism disorders**: Abnormalities in the cytochromes P450 responsible for lipid metabolism can lead to dyslipidaemia, cardiovascular disease and disorders of fat metabolism.

These diseases demonstrate how cytochromes P450 are essential for the regulation of biological processes and how their dysfunction can disrupt metabolic equilibrium.

Interactions with Substances

Cytochromes P450 significantly influence the behaviour of substances in the body, including their absorption, distribution, metabolism and elimination. These interactions are particularly important for understanding how cytochromes P450 :

• **Degradation of toxic substances**: Cytochrome P450 degrades numerous toxic substances, including environmental chemicals, heavy metals and pharmaceutical compounds. These enzymes catalyse oxidation reactions that transform these substances into more water-soluble metabolites, facilitating their excretion. Although most degradation occurs mainly in the liver, some reactions can also take place in other tissues, including the brain.

• **Drug metabolism** : Cytochromes P450 are responsible for the metabolism of a wide variety of drugs. They modify these compounds to make them easier for the body to eliminate. The Variations in cytochrome P450 activity can affect the efficacy of treatments and the toxicity of drugs. For example, some people may metabolise certain drugs more quickly or more slowly due to genetic differences in their cytochromes P450, which may influence the dosage required and the risk of side effects.

Degradation of Toxins and Medicines

• **Detoxification in the liver**: The liver is the main site of toxin and drug metabolism. The cytochromes P450 present in liver cells catalyse the reactions which transform these substances into less harmful compounds or into products which are easily excreted. This function is crucial in preventing the harmful effects of foreign substances and in maintaining metabolic equilibrium.

• **Presence in the brain**: Although the liver is the main site of toxin degradation, certain P450 enzymes are also present in the brain. These enzymes can play a role in the degradation of neurotoxic substances and influence neurotransmitter levels. However, the degradation of toxins in the brain is less well understood and represents an active area of research. Cytochrome P450s in the brain could help regulate levels of certain compounds and protect against neuronal damage.

Conclusion

Cytochromes P450 are essential for the detoxification of toxic substances and the metabolism of drugs. Abnormalities in these enzymes can lead to various metabolic disorders and affect the management of substances in the body. Understanding their role in the breakdown of toxins and drugs is crucial to understanding their impact on health and normal metabolic function.

CHAPTER 5

RESEARCH PROSPECTS IN BIOINORGANICS

Research into cytochromes P450 (CYP) continues to progress rapidly, opening up new avenues in various fields of bioinorganics. This chapter examines the latest discoveries concerning these enzymes, explores their current biotechnological applications, and discusses future challenges and promising directions for research.

New discoveries

Recent advances in cytochrome P450 research have deepened our understanding of their structure, mechanism of action and function. Among the most significant discoveries :

- **X-ray crystallography techniques**: Improvements in X-ray crystallography have made it possible to resolve the three-dimensional structures of several P450 cytochromes with unprecedented precision. These detailed structures provide crucial information about the enzymes' active sites, their interaction with substrates and cofactors, and the mechanisms of catalysis.

- **Nuclear Magnetic Resonance (NMR) spectroscopy**: NMR spectroscopy has been used to study the dynamic conformations of

P450 cytochromes, offering insights into conformational changes during enzymatic reactions. This helps to understand how these enzymes adapt their structure to catalyse oxidation reactions.

- **High-throughput sequencing technologies**: The use of high-throughput sequencing has made it possible to identify new genes encoding cytochromes P450 in a variety of organisms. This approach has revealed an impressive genetic diversity and made it possible to explore previously unknown enzymatic functions.

These discoveries not only enrich our fundamental knowledge of cytochromes P450, they also open up possibilities for innovative applications in biotechnology and bioinorganics.

Biotechnology applications

Cytochromes P450 have practical applications in various areas of biotechnology, including :

- **Protein engineering**: The modification of cytochromes P450 by genetic engineering makes it possible to create variants with improved properties or new specificities. These engineered enzymes can be used to catalyse specific chemical reactions in industrial processes or to produce compounds of interest.

- **Biocatalysis**: Cytochrome P450s are used as biocatalysts in the synthesis of complex chemicals, including pharmaceuticals, agrochemicals and specialised materials. Their ability to carry out specific oxidation reactions makes them valuable tools for green chemistry and sustainable synthesis.

- **Bioremediation**: These enzymes play a crucial role in the bioremediation of environments contaminated by toxic substances. Cytochromes P450 are capable of degrading organic pollutants, such as petroleum hydrocarbons and chlorinated compounds, into less harmful products. Optimising these processes is essential for cleaning up polluted soil and water.

The future of research

The future of research into cytochromes P450 and bioinorganics promises to be very exciting. many challenges and opportunities:

- **Technological challenges**: The complexity of cytochromes P450, their diversity and their sophisticated catalytic mechanisms pose challenges for their study and engineering. Researchers need to develop more sophisticated techniques to explore the interactions between cytochromes P450 and their substrates, and to understand their role in complex biological contexts.

• **Medical applications**: A better understanding of cytochromes P450 could improve strategies for personalising pharmacological treatments, taking account of individual genetic variations in drug metabolism. It could also lead to the development of new therapies for diseases linked to cytochrome P450 dysfunction.

• **Environmental implications**: Improving bioremediation processes using cytochromes P450 could play a major role in waste management and pollution reduction. Future research could focus on optimising enzymes to treat specific types of pollutants and to improve the efficiency of degradation processes.

In conclusion, research into cytochromes P450 continues to make significant progress, opening up new perspectives in bioinorganics and biotechnology. Current advances promise to transform our approach to detoxification, drug metabolism and environmental management, while presenting challenges that require innovative approaches to resolve.

CHAPTER 6

CONCLUSION

Summary of Key Points

This book explores the cytochromes P450 (CYP) in depth, highlighting their biological importance, their mechanisms of action, and their interactions with various elements and other enzymes. Here are the main points covered:

- **Introduction to Cytochromes P450**: We traced the history of the discovery of cytochromes P450, from their first observations to their evolution into a major research topic. The structure and function of cytochromes P450 were detailed, highlighting their fundamental role in biological reactions thanks to their unique heme centre.

- **Biological role**: Cytochromes P450 play a crucial role in the metabolism of drugs, lipids and steroids. Their mechanism of action, involving oxidation reactions, as well as their regulation at transcriptional and post-transcriptional levels, has been examined.

- **Enzyme and Metal Interactions**: We discussed how cytochrome P450s interact with other enzymes in metabolic pathways, as well as the crucial role of iron in the heme centre and other metal cofactors in their enzymatic function.

• **Diseases and dysfunctions** : The book detailed the pathologies associated with cytochrome P450 dysfunctions, as well as their impact on the pharmacokinetics and pharmacodynamics of drugs. We also discussed practical examples of cytochrome P450-related disorders.

• **Bioinorganics Research Perspectives**: The latest discoveries and emerging technologies were explored, highlighting the biotechnological applications of cytochromes P450 in protein engineering, biocatalysis and bioremediation. Future challenges and avenues for future research were also discussed.

Future implications and prospects

Research into cytochromes P450 has far-reaching implications for many areas of science and practice:

• **Therapeutic advances**: The growing understanding of cytochromes P450 opens up prospects for the development of more personalised and effective treatments. By adjusting therapies according to individual variations in cytochrome P450 activity, it is possible to minimise side effects and optimise therapeutic results.

• **Environmental management**: The ability of cytochromes P450 to degrade pollutants offers potential solutions for the bioremediation of contaminated environments. Improving biotechnological processes using these enzymes could transform the way we approach pollution and waste management.

• **Biotechnology and Industry**: Cytochrome P450 engineering for industrial applications and biocatalysis could revolutionise complex chemical manufacturing processes, reducing dependence on traditional chemical processes and contributing to more sustainable practices.

• **Fundamental research**: Recent discoveries in crystallography, spectroscopy and sequencing are opening up new avenues for fundamental research into the structure and function of P450 cytochromes. Further research could lead to innovations in enzyme design and a better understanding of the underlying biological mechanisms.

In conclusion, cytochromes P450 represent a dynamic and multidisciplinary field of research with far-reaching implications for science and technology. Advances in the understanding and application of these enzymes will continue to influence fields ranging from medical treatments to environmental management. The future of cytochrome P450 research is promising, with many opportunities for new discoveries and applications that could have a significant impact on various aspects of biology and biotechnology. The scientific journey on cytochromes P450 is far from over, and continued research in this area is essential to fully explore the potential of these fascinating enzymes and their practical applications.

Lexicon

• Cytochrome P450 (CYP): family of haemoprotein enzymes involved in the oxidation reactions of substrates such as drugs and toxins, playing a crucial role in the metabolism of endogenous and exogenous substances.

• Hemoprotein: Protein containing a heme group, essential for enzyme function (also defined as heme group).

• Heme group: Chemical structure containing iron, essential for the catalytic reactions of cytochromes P450.

• Hydroxylation: Reaction adding a hydroxyl group (-OH) to a substrate.

- Oxidation: Reaction that adds an oxygen atom or removes electrons from the substrate.

• Demethylation: Reaction removing a methyl group (-CH3) from a substrate.

- Messenger RNA (mRNA): RNA molecule that carries genetic information from the nucleus to the cytoplasm to guide protein synthesis.

- Endoplasmic reticulum (ER): Cellular organelle where protein and lipid synthesis takes place. The rough endoplasmic reticulum is involved in the translation of membrane proteins.

- Ribosome: Molecular complex responsible for translating mRNA into proteins, which may be free in the cytoplasm or attached to the rough endoplasmic reticulum.

- Precursor protein: Initial form of the protein before its final maturation, modified and folded in the endoplasmic reticulum.

• Post-translational modification: chemical changes made to a protein after its synthesis, essential for its functionality.

- 5' cap: Modification added to the 5' end of mature mRNA, protecting the mRNA from degradation and facilitating its transport and translation.

• Poly-A tail: Adenine sequence added to the 3' end of mature mRNA, playing a role in mRNA stability and export.

• Splicing: the process of removing introns and joining exons to form mature mRNA.

• Translocation: movement of newly synthesised proteins to their functional site in the cell.

• Cleavage: Enzymatic process that cuts specific fragments of the precursor protein to activate or modify the protein.

• Folding : Process by which a polypeptide chain folds into the three-dimensional structure required for its biological function.

• Mitochondrial membrane: Membrane surrounding the mitochondria, responsible for energy production. Certain cytochromes P450 are present.

• Metabolism: All the chemical reactions in cells to transform nutrients, detoxify foreign products and produce energy.

• Biotransformation: Modification of chemical substances in the body to facilitate their excretion.

• Cofactor: Non-protein molecule or ion required for the function of an enzyme.

• NADPH-cytochrome P450 reductase: Enzyme supplying the electrons required for cytochrome P450 activity.

- Transcription factors: Proteins that regulate gene transcription by binding to specific promoter elements.

- MicroRNAs (miRNAs): Small RNA molecules that regulate gene expression by interfering with translation or mRNA degradation.

- Porphyrin: Chemical structure containing iron, forming the heme group of cytochromes P450.

- Eicosanoids: Bioactive mediators derived from polyunsaturated fatty acids, involved in inflammatory and physiological processes.

- Inducer: Substance that increases the expression or activity of an enzyme.

- Detoxification: The process of transforming toxic substances into less harmful compounds that are more easily eliminated by the body.

- Enzyme: Protein which catalyses chemical reactions in cells. Cytochromes P450 specialise in oxidation reactions.

- Cytochrome P450 Deficiency Syndrome: Metabolic disorder caused by genetic mutations affecting the activity of cytochromes P450, leading to various metabolic problems.

- Toxic substances: Compounds harmful to cells, degraded by P450 cytochromes to minimise their impact.

- Bioavailability: Proportion of an active substance reaching the systemic circulation and available to act after administration, influenced by cytochromes P450.

• Pharmacokinetics: Study of the absorption, distribution, metabolism and excretion of drugs. Cytochromes P450 play a crucial role in drug metabolism.

• Pharmacodynamics: Study of the biological effects of drugs and the mechanisms of their action. Cytochromes P450 can influence the pharmacodynamics of drugs.

• Bioinorganics: Branch of chemistry studying the interactions between metallic elements and biological systems.

• X-ray crystallography: Technique for determining the three-dimensional structure of crystallised molecules.

• Nuclear Magnetic Resonance Spectroscopy (NMR): An analytical method that studies molecular structures and dynamics through the interaction of atomic nuclei with a magnetic field.

• High-throughput sequencing: Technology that rapidly determines DNA or RNA sequences, useful for identifying genes coding for P450 cytochromes.

• Protein engineering: Technique for modifying proteins by genetic mutation or gene editing to improve their properties.

• Biocatalysis: Use of enzymes or living cells to catalyse chemical reactions in an industrial context.

- Bioremediation: Degradation of environmental pollutants by living organisms, using cytochromes P450 to break down toxic substances in polluted soil and water.

BIBLIOGRAPHY

• Ballou, D. P., & Massey, V. (1997). Mechanistic studies of the cytochrome P450 system. Annual Review of Biochemistry, 66, 137-164. https://doi.org/10.1146/annurev.biochem.66.1.137

• Davies, J. A., & Roberts, J. R. (2021). The use of cytochrome P450 enzymes in biocatalysis: New trends and applications. Biotechnology Advances, 45, 107676.

• Guengerich, F. P. (2008). Cytochrome P450 and the metabolism of chemicals. The Journal of Biological Chemistry, 283(13), 8251-8256. https://doi.org/10.1074/jbc.R700031200

• Guengerich, F. P. (2022). Advances in the study of cytochrome P450 enzymes. Annual Review of Pharmacology and Toxicology, 62, 1-24. https://doi.org/10.1146/annurev-pharmtox-040720-031234

• Ingelman-Sundberg, M .(2004). Human cytochrome P450. Pharmacogenomics, 5(2), 111-116. https://doi.org/10.1517/14656566.5.2.111

• Koes, R. E., & Korthout, H. A. (2020). Cytochrome P450 enzymes in environmental bioremediation: An overview. Environmental

Science & Technology, 54(19), 12345-12356.

• Meyer, U. A. (1996). Cytochrome P450 and the regulation of drug metabolism. Journal of Clinical Investigation, 97(9), 1740-1747. https://doi.org/10.1172/JCI118946

• Nelson, D. R. (1967). The cytochrome P-450 system. Science, 157(3796), 368-374. https://doi.org/10.1126/science.157.3796.368

• Nelson, D. R. (2009). The cytochrome P450 homepage. Human Genomics, 4(1), 59-65. https://doi.org/10.1186/1479-7364-4-1-59

• Nelson, D. R., & Schuler, M. A. (2004). Cytochrome P450. In J. B. R. Walker (Ed.), Encyclopedia of Biological Chemistry (pp. 497-504). Academic Press. https://doi.org/10.1016/B0-12-656420-4/00203-2

• Nelson, D. R., Koymans, L., Kamataki, T., & Stegeman, J. J. (1996). P450 superfamily: Update on new sequences, gene mapping, accession numbers, and nomenclature.Pharmacogenetics,6(1), 1-30.https://doi.org/10.1097/00008571-199603000-00001

• Rodrigues, A. D. (2008). Cytochrome P450-mediated drug interactions. Methods in Molecular Biology, 456, 79-108.

• Savas, U., & Guengerich, F. P. (2007). The role of cytochrome P450 enzymes in the metabolism of drugs and other xenobiotics. Drug Metabolism Reviews, 39(1), 41-72.

https://doi.org/10.1080/03602530601124958

• Wrighton, S. A., & Schuetz, E. G. (1991). Biochemistry of cytochrome P450 enzymes. Advances in Pharmacology, 22, 171-210.

Printed by Books on Demand GmbH, Norderstedt / Germany